Name ______________________________ Date ________

Nature of Light

Light is a form of energy we call **radiant energy**. The light we can see is called visible light, but there are other kinds of radiant energy, such as radio waves and x-rays, which cannot be seen by the human eye. A **ray** is a beam of light energy.

Scientists think that light rays travel through space in waves. The waves spread out into space. The brighter the light, the farther the waves travel. The bright light from our sun travels 150 million kilometers (93,000,000 mi.) to Earth and beyond.

Light travels very fast. In space it travels about 186,000 mi. per second, a distance equal to about seven and one half trips around the Earth. Even at this speed, it takes eight minutes for sunlight to reach the earth.

Light travels at different speeds as it passes through different materials. Water is thicker than air, so light travel only three-fourths as fast in water as it does in air.

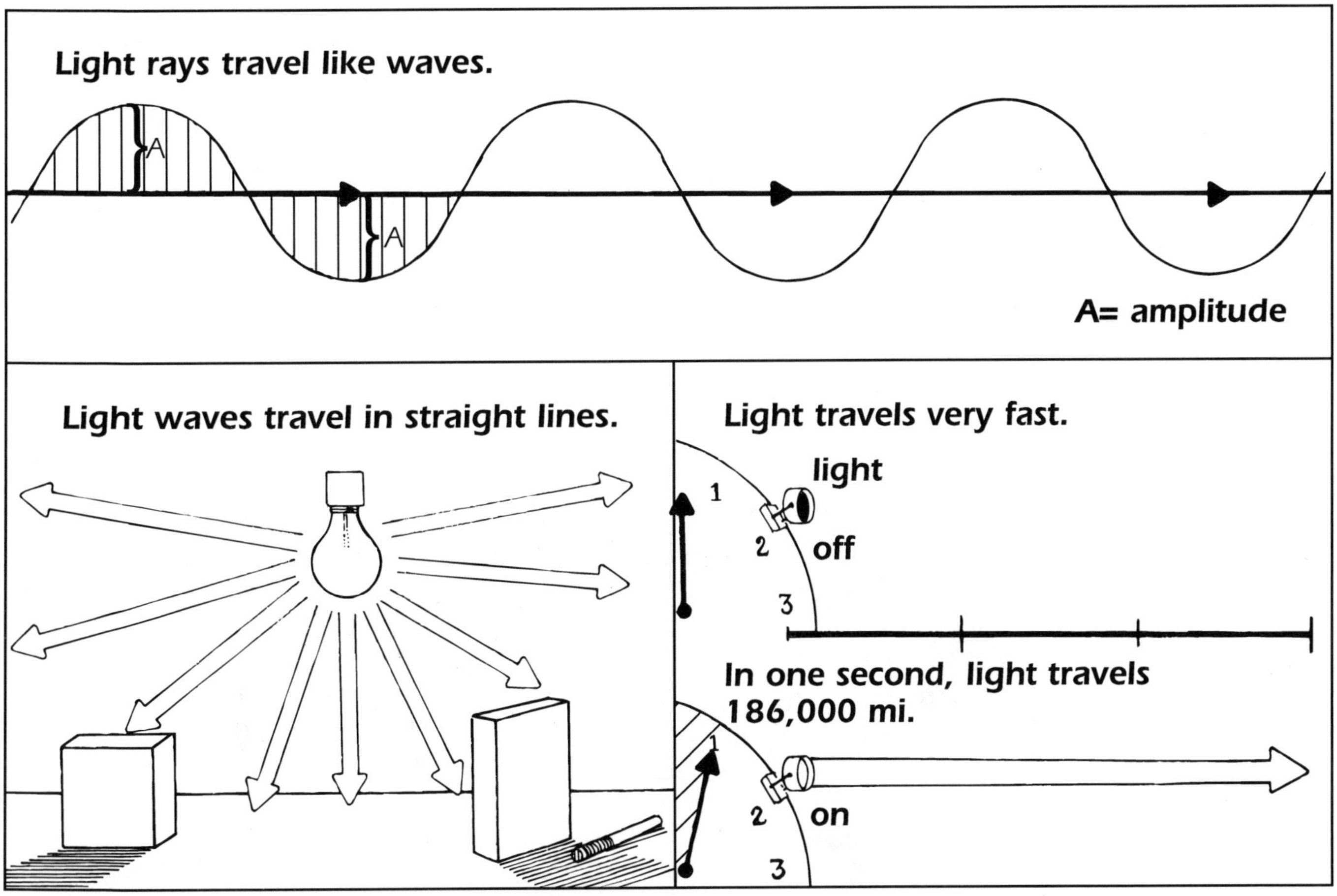

1. Light rays travel like ______________________________.
2. Circle the picture that uses electricity to make light.
3. It takes one ____________ for light to travel 186,000 miles.
4. Light waves travel in ____________ lines.

Name ______________________ Date ______________

LIGHT & SOUND

Reflection

Light waves move away from their source in straight lines. They stop when they hit something they cannot go through, or they may bounce back as **reflected** light. When an object does not reflect light, it is because its surface is absorbing it. Some things reflect light better than others. Smooth, shiny surfaces like mirrors reflect light well; dark, rough surfaces do not.

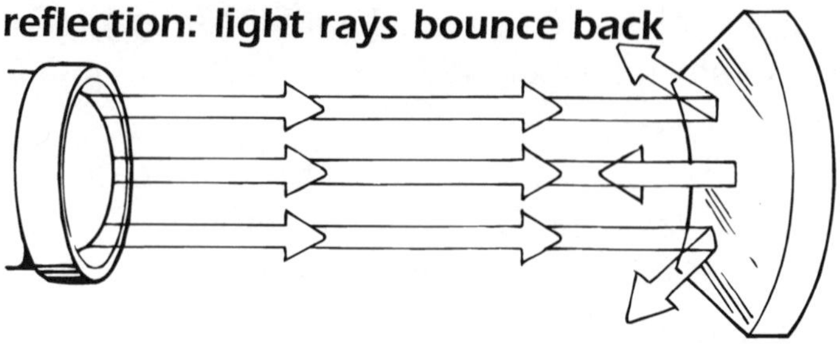

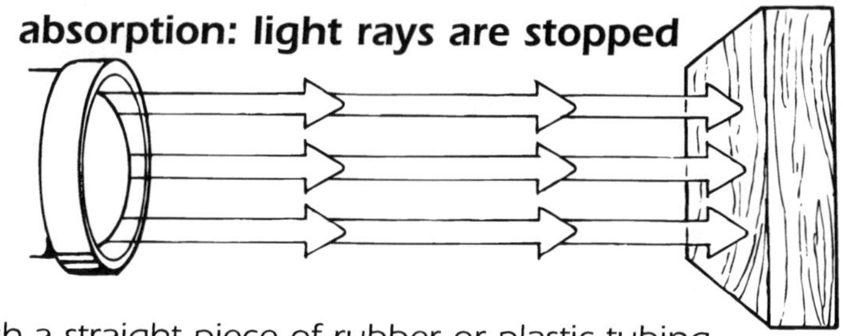

Activity

To see how light moves, look at a light through a straight piece of rubber or plastic tubing. You can also use a sheet of paper rolled into a tube. (Do not look directly at a strong light that hurts your eyes.) Bend the tube a little so that it is not straight.

1. Can you still see the light? Why or why not? ______________________

2. When a solid object stops light rays, a shadow is formed behind it. Draw the tree's shadow in the left drawing.
3. Would the tree in the right drawing make a shadow? ______________
4. Why? ______________________

Name ______________________ Date ______________

Refraction

Light travels in straight lines, but it can be **refracted**, or bent. Refraction occurs only when the light rays enter or leave a substance at an angle. If the rays are straight up and down, they do not bend.

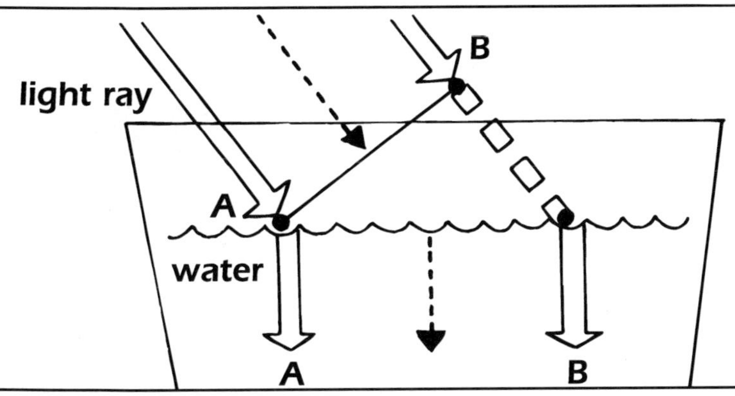

refraction: bending of light rays

A hits the water first.

B travels faster (dotted line) until it hits water.

Then A and B travel at the same speed, but the direction of the light ray has changed.

Activity

Fill a clear glass half full with water. Put a pencil into the glass. Look at the pencil through the side of the glass.

How does the pencil look different than it did before you put it in the water?

____

Draw the way it looks in the glass at the right.

Light travels faster in some materials than in others, and some materials refract light to a greater degree than others. When light rays pass from one material to another, they may speed up, slow down, or change direction.

To see how this happens, roll a pencil down a sloping surface. Slow down one end of the pencil by placing an obstacle in its way. The other end of the pencil is not slowed, so it swings around. The direction of the pencil changes.

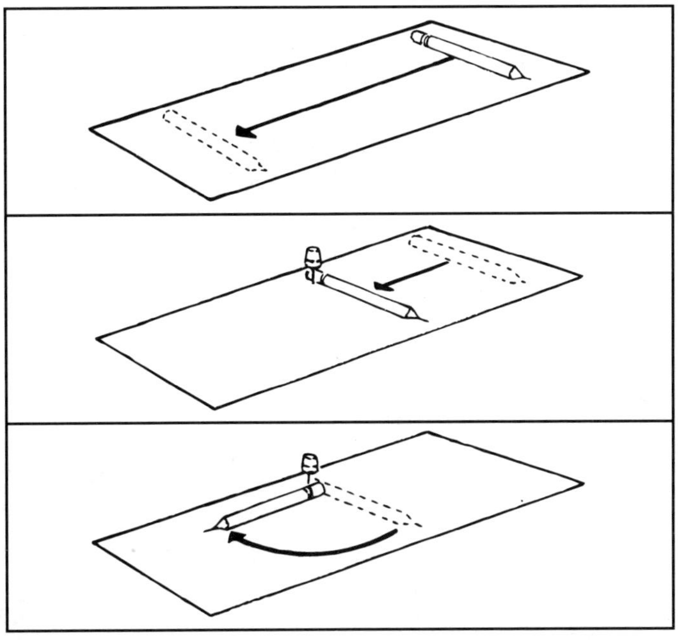

Name ________________________________ Date ____________________

LIGHT, MATERIALS, AND SHADOWS

If an object lets light pass through it, as a clear pane of glass does, it is called **transparent**. If an object lets only some light pass through, the way frosted glass does, it is called **translucent**. If an object lets no light pass through, it is **opaque**. Opaque objects create shadows.

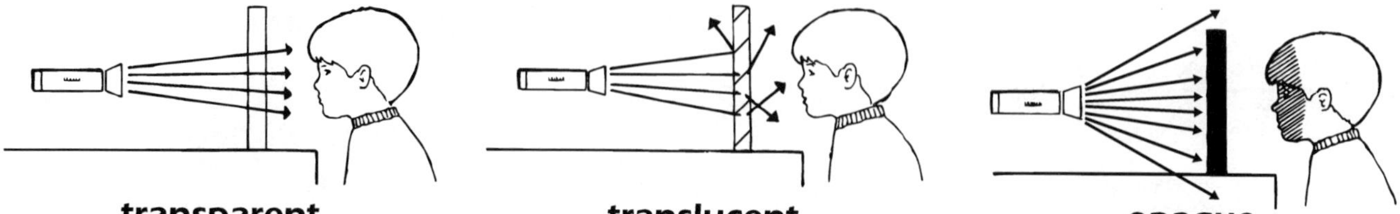

transparent **translucent** **opaque**

ACTIVITY

Hold different materials (suggested materials: waxed paper, aluminum foil, paper, cloth, cardboard, water, wood, air, and your own hand) up to a flashlight or light bulb. Decide which materials are transparent, which are translucent, and which are opaque. Write the names of the materials in the correct boxes.

transparent	translucent	opaque

ACTIVITY

Place a sheet of white paper or cardboard behind the opaque objects. Darken the room a little. Shine the light on each object and notice the shadow it makes on the white surface. Try changing the position of the light and the object until you are able to change the size of the shadow.

1. What makes the shadow get larger? ________________________________

2. What makes it get smaller? ________________________________

Shadows outdoors change their shapes throughout the day as the direction of the sun's rays changes.

Name ______________________ Date ______________

Light and Color

The white light that is visible to us is really a mixture of many colors. These colors can be seen when a beam of light passes at an angle through a glass **prism**. When the light passes from the air through the prism, the rays change direction. They are refracted. The different colors are bent differently. Violet is bent the most and red the least. When the light comes out of the prism, the different colors travel in slightly different ways and hit at different places, forming connecting bands of color like a rainbow, or a **spectrum**.

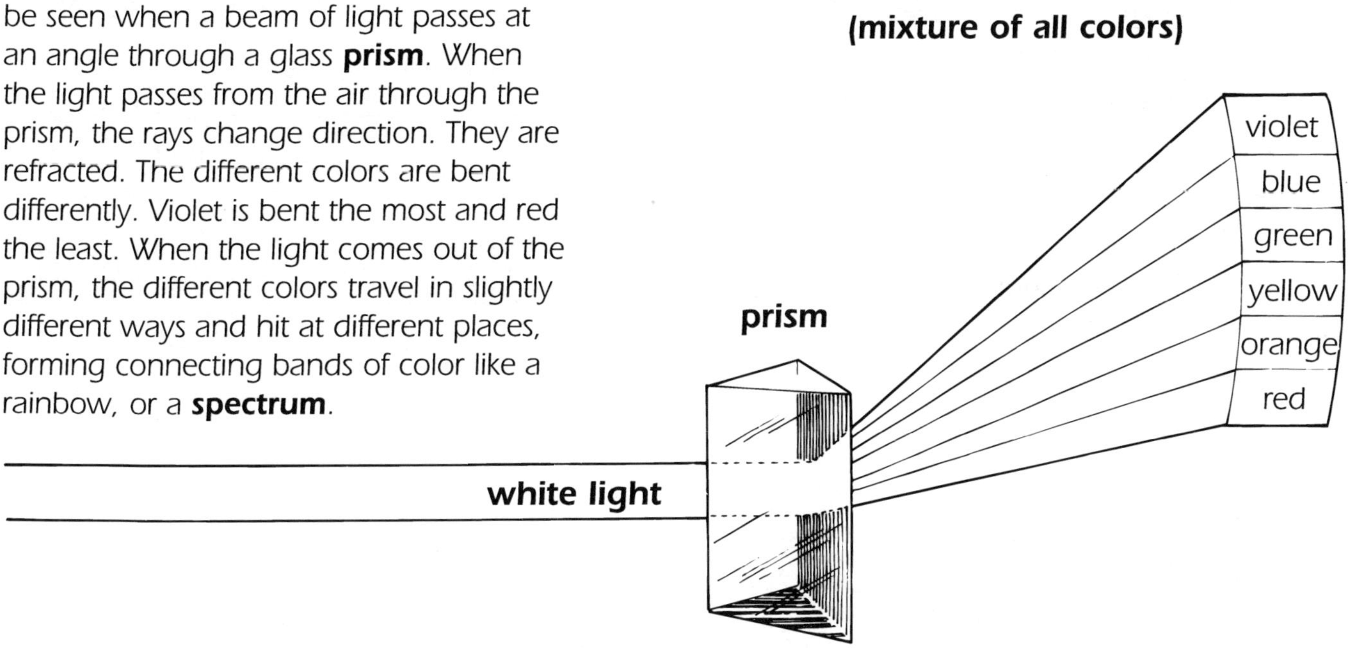

Activity

Produce a spectrum by holding a prism near a window that is being hit directly by rays of the sun. (If you do not have a prism, a glass of water will often give the same results.) Allow the sunlight to pass through the prism. Move and twist the prism around in the sunlight until you see a rainbow (spectrum) on the ceiling, floor, or wall. Place white paper behind the "rainbow" so you can see the colors clearly.

1. List the colors you see in order, starting with red. Six colors may be visible.

____

What gives objects their color? Objects can reflect only the light that strikes them. A red shirt reflects the part of light that we see as red. It absorbs the other colors of the spectrum.

2. What color is an object that does not reflect any light? ______________
3. What colors are white paper reflecting? ______________
4. Dark objects absorb ______________ light than do light-colored objects.

Not all light contains all the colors in the same amounts as sunlight does. A yellow light, such as a "bug light," has much less red, blue, and green than white light does. Since an object's color comes from the light it reflects, objects have different colors under different lights.

Name ______________________ Date ______________

LIGHT & SOUND

BEHAVIOR OF LIGHT REVIEW

Rearrange the letters to spell the words that belong in the spaces. Then find the words in the puzzle.

1. Light is a form of ______________ energy. (TARDIAN)
2. Light travels in ______________ lines. (THIRTAGS)
3. An object that lets light pass through it is ______________. (SANTERPRANT)
4. When light rays are bent, we say they are ______________. (CREAFDERT)
5. If an object lets no light pass through it, it is ______________. (EPOQUA)
6. A rainbow of colors is called a ______________. (MECTURPS)
7. When light rays bounce back, we say they are ______________. (FLERTDEEC)
8. ______________ is the color of the spectrum that is bent the most. (TOIVEL)
9. Light we can see is called ______________ light. (BLIEVIS)
10. A dark color ______________ more light than a light color. (BASROBS)

J	V	D	O	L	P	A	B	S	O	R	B	S	W
L	I	S	X	U	M	T	C	U	F	G	I	P	O
V	E	T	N	Y	Z	R	H	V	I	O	L	E	T
I	C	R	A	D	I	A	N	T	D	I	J	C	E
S	M	A	K	L	O	N	P	R	S	U	V	T	W
I	X	I	Z	E	U	S	F	O	M	L	U	R	Q
B	O	G	H	A	O	P	A	Q	U	E	M	U	U
L	I	H	P	R	U	A	T	U	F	M	I	M	I
E	N	T	A	G	W	R	W	S	O	X	H	Z	L
Z	L	J	H	O	R	E	F	R	A	C	T	E	D
O	M	P	S	U	G	N	S	U	T	E	D	P	T
R	E	F	L	E	C	T	E	D	Y	K	L	G	O

Name ______________________ Date ______________

LENSES

A **lens** is a piece of glass or other clear material that has smoothly curved sides. Refraction causes light to bend when it passes through a lens. There are two basic kinds of lenses: **convex** and **concave**.

A convex lens is thicker in the middle than at the edges. A magnifying glass is a convex lens. If it is held close to an object, it will make it look larger. Light rays passing through a convex lens bend inward so that they meet at a point (the focus) on the other side. The **focus** is the point where the image of the object is formed.

With convex lenses, the light rays cross and come out of the lens on opposite sides. This makes images that appear turned around or upside down.

ACTIVITY

Draw an arrow on a card and prop it up on a table. Place a glass of water a little to one side of the card and a few inches in front of it. Now stoop down until your eyes are on a level with the arrow. Notice which way the arrow points, and then slide the glass until it is between the card and your eyes.

1. Which way is the arrow pointing now? ______________________

2. Slide the glass of water in and out a few times and watch the arrow. What happens?

____

3. What causes this? ______________________

Lenses that are thicker at the edges are called concave. Light rays passing through a concave lens bend outward, so they spread apart. The images produced are right-side-up and smaller than the object.

Write the name of each lens on the lines under the diagrams.

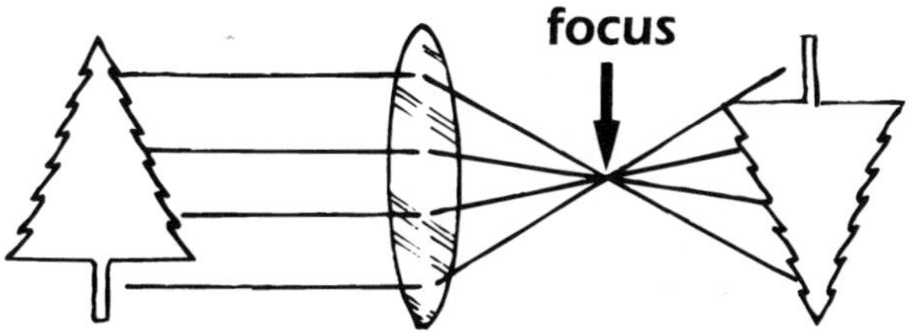

4. ______________________

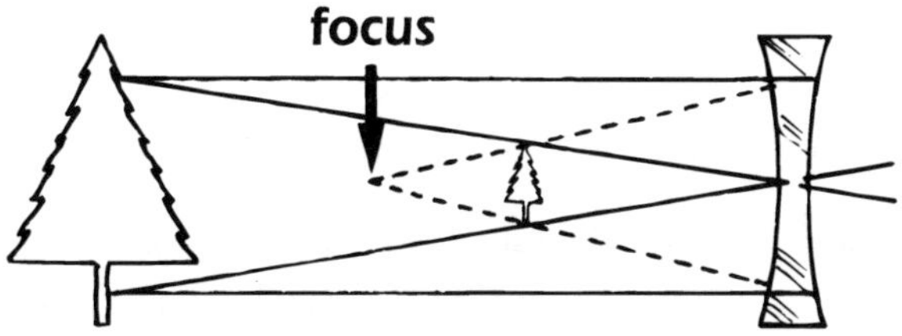

5. ______________________

Name ______________________ Date ______________

LIGHT & SOUND

LIGHT AND EYESIGHT

The eyeball is protected by a clear, outer layer called the **cornea**. The colored part of the eye is called the **iris**. In the center of the iris there is a black spot called the **pupil**. The pupil is an opening in the iris. Light gets into your eyes through the pupil. The pupil looks dark because no light reflects off the back of the eye.

How we see

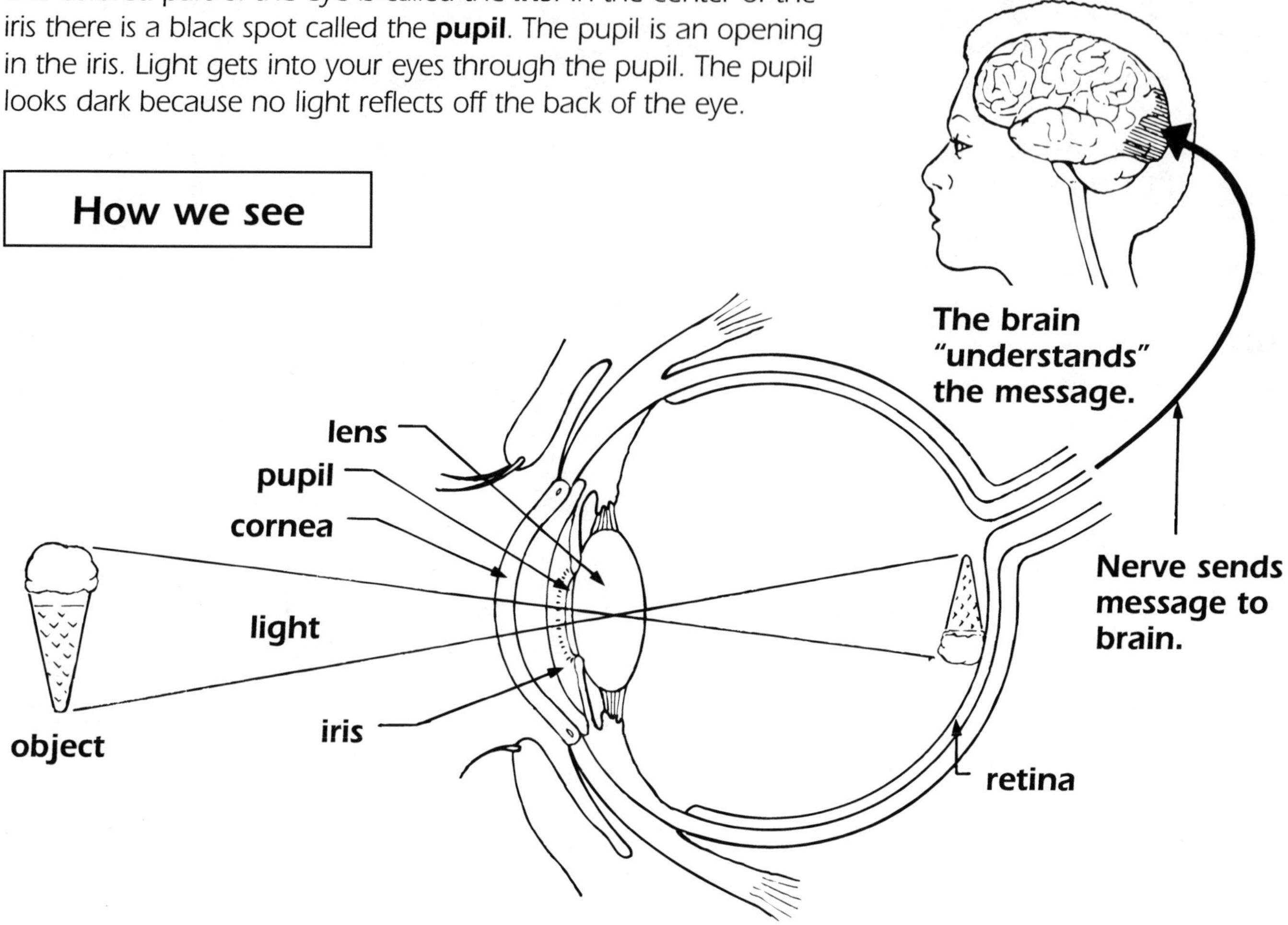

ACTIVITY

Look at a friend's eyes while he or she stands in a brightly-lit area, or looks at something bright. (CAUTION: Do not look directly at the sun or at any light that hurts your eyes.) Notice the size of the pupils in your friend's eyes. Quickly go to a darker place.

1. Now what happens? ______________________

2. Why does this happen? ______________________

Just behind the pupil is a **lens**. It is shaped somewhat like a magnifying glass (convex). When you look at something, the lenses in your eyes focus the picture (image) on the back of your eyeball, on the **retina**. The picture is upside down. Nerves in the back of your eyes take the image to your brain. Your brain then makes you see the picture right side up.

Name ______________________________ Date ____________________

A Camera Is Like an Eye

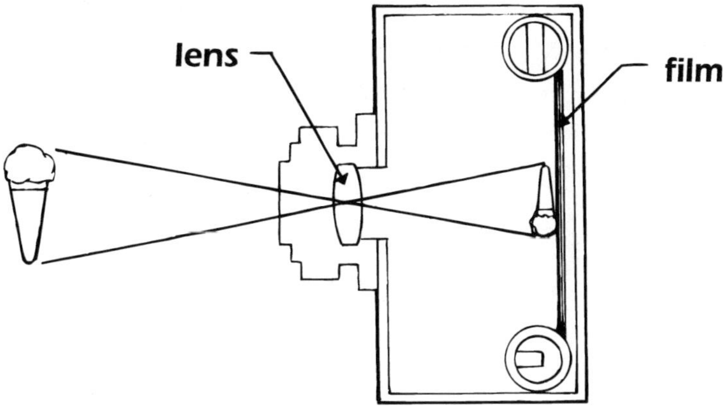

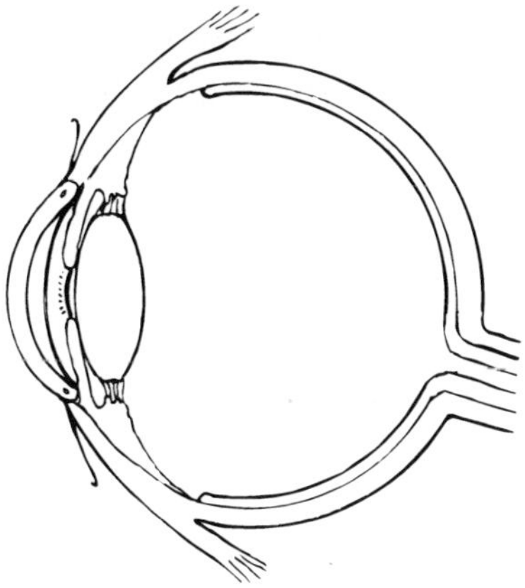

1. What part of the eye and the camera have the same name? ______________
2. The retina of the eye is like the ______________ of the camera.
3. What part of the head tells us what the eye sees? ______________
4. Does the camera have a nerve? ______________
5. What do we call the upside-down picture on the retina? ______________

As we saw before, the size of the pupil controls the amount of light entering the eye. The **aperture** (**a**'per cher) is the opening in the camera that is like the pupil.

Label the parts of the eye in the drawing at the top of the page.
Below draw the eye as it would look in a dark room.

To get a clear photograph, we adjust the camera lens to focus the image on the film. The lens of your eye also becomes thicker or thinner, according to the distance of the object. It will thicken for close objects. But if objects are far away, the light doesn't have to be bent so much. The muscles can then relax, and the lens gets thinner.

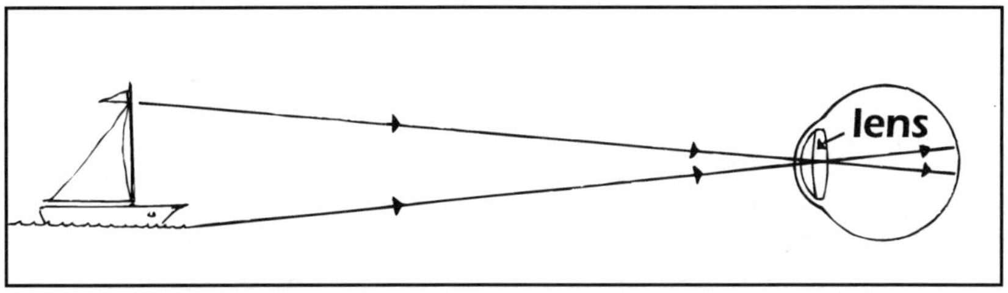

Name ____________________ Date ____________

Law of Reflection

Anything that glows sends out waves of light—the sun, a firefly, a candle flame, a light bulb—but very few objects that we see send out their own light waves. Most things are seen by the reflection of light. For example, the sun shines on a house, and the waves bounce off to your eyes so you see it. At night, if the lights are turned off, there are no light waves to be reflected, and you cannot see anything.

Light reflects off an object at an angle equal to the angle at which it first strikes it. This is best seen when looking in a mirror.

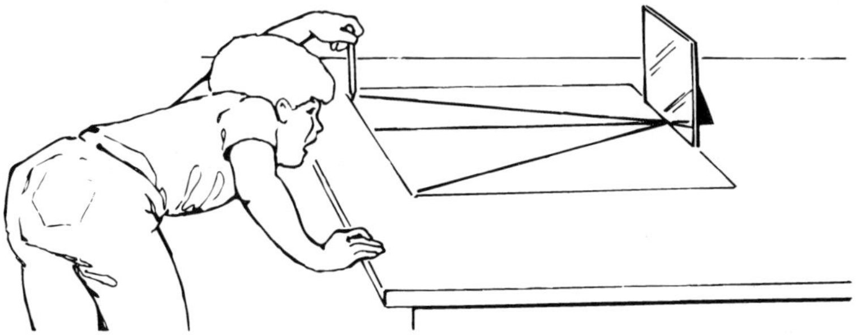

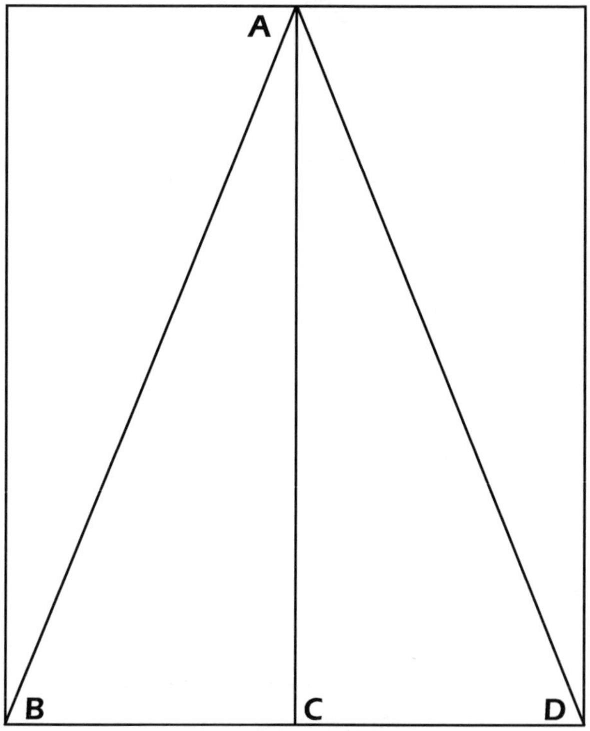

Activity

Make a diagram on a large sheet of paper like the one shown at the right. Mark point A at the middle of the upper edge of the paper. Mark point C at the middle of the lower edge of the paper. Mark points B and D at the lower corners. Draw straight lines between the points as shown.

Hold a mirror along the upper edge of the paper at point A. Look along the line from point C to point A.

1. How do you know that light traveled in a straight line when you looked along line C–A?

   ____

2. Put your pencil on point D and look along the line from points B to A. Can you see the pencil?

   __________ Explain why. ____________________

3. Can you see the pencil if you put it on point C? ____________________

   Explain why. ____________________

Name ______________________________ Date ______________

SOURCES OF LIGHT

Light comes to our eyes in two ways. Light from the sun or a light bulb comes directly to our eyes. Objects like these that make their own light are called **luminous**. Objects that do not produce their own light may be seen only when light from some source hits them and bounces back to our eyes as reflected light. Light is produced in many different ways. When light comes from some human activity, as in an electric light bulb or candle, it is called **artificial light**. When light comes from something in nature, we call it **natural light**.

natural light

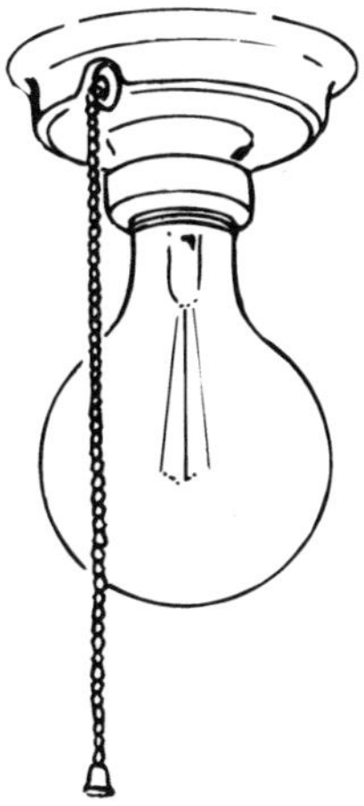

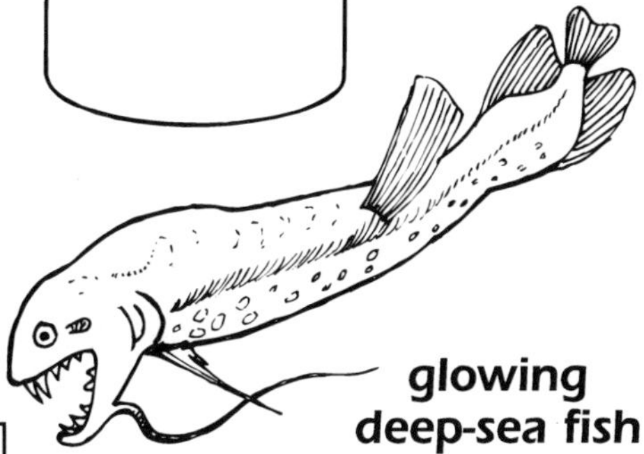

glowing deep-sea fish

artificial light

1. In the pictures above, draw a line from the picture to the box that shows where the light is made.
2. Circle the picture of the insect that makes its own light.
3. Two ways light can be made are ______________ and ______________.
4. Which lights are luminous? ______________________________
5. Which lights are artificial? ______________________________

Name ______________________ Date ______________

ENERGY FROM SUNLIGHT

Every living thing on our planet depends on sunlight. Plants cannot grow without light. The leaves of green plants trap energy from the sunlight. When animals eat the plants, this energy from the sunlight is passed on.

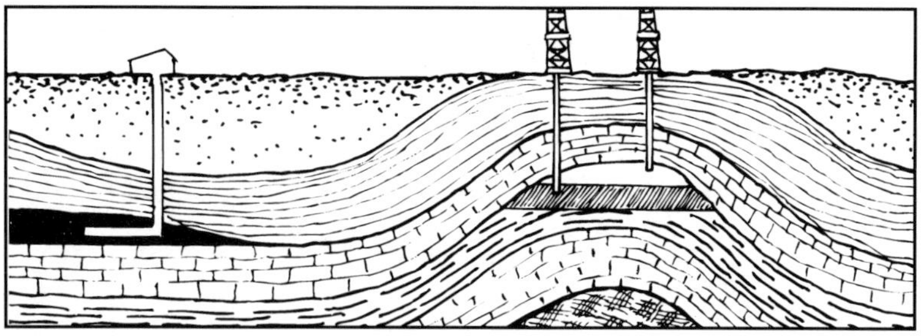

Fuel is stored energy from the sun. Gasoline used in automobiles comes from oil. Oil was formed by rotting plants when Earth was much younger. Coal was also made from plants of long ago. It burns to make electricity and is used in making plastic.

Sunlight is a part of Earth's water cycle. The heat energy in sunlight evaporates water and carries it into the air. Eventually, the water returns as rain or snow.

The ultraviolet rays in sunlight help the body produce vitamin D, which makes strong teeth and bones. (Too much exposure to sunlight, however, harms the body.)

1. Name three important uses of sunlight: ______________________

____

2. Where do coal and oil come from? ______________________
3. What part of green plants traps the sun's energy? ______________________
4. Some animals eat meat for food. Where does the energy in meat come from?

____

Name ______________________ Date ______________

LIGHT & SOUND

Light in Technology

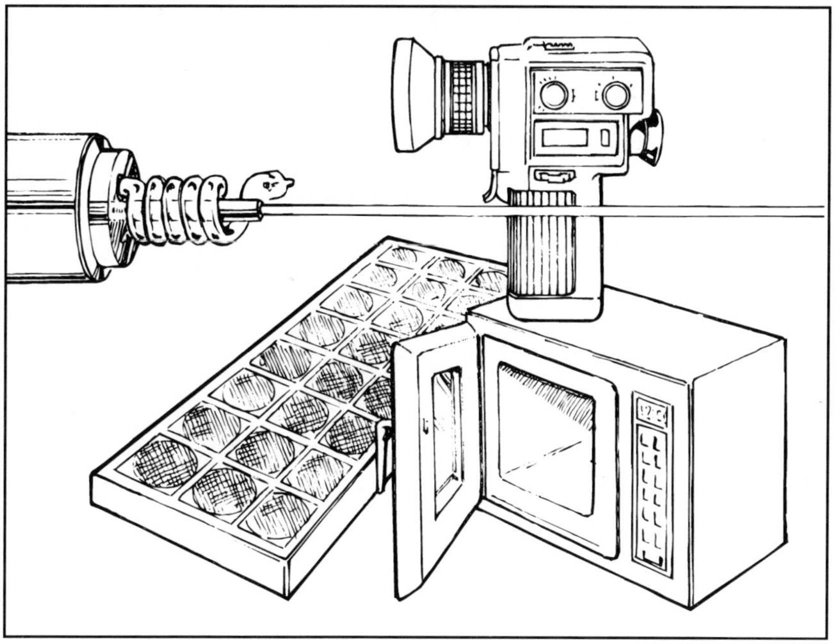

Scientists have found many uses for light in technology. When we listen to radio or watch television, we are using the light energy known as **radio waves**. The **microwaves** in a microwave oven are light waves, too.

Light energy can be changed into electrical energy by using a **solar cell**. The cell has the chemical silicon and other chemicals in it. When sunlight hits the silicon, a small electric current flows. A group of solar cells connected to each other is called a solar battery. Solar batteries have been used in satellites which orbit earth. Solar houses are heated by stored energy from sunlight.

Light can change and react with various chemicals. This is what happens in photography when film is exposed (to light) and developed. Certain chemicals give off small electric currents when struck by light. This is what produces the "electric eyes" used in automatic doors.

The **laser** is a device that sends out a narrow beam of intense light in only one direction. Lasers have many uses: welding metals, piercing hard substances, treating eye problems, and transmitting voice and television signals.

1. Name five important technological uses of light. ______________________

____

2. What does a solar cell do? ______________________

   Use reference sources: ______________________

3. The word laser is an acronym. (That means each letter stands for a word.) What does L-A-S-E-R stand for? ______________________

____

4. Radar, another example of light in technology, is also an acronym. What does R-A-D-A-R stand for?

____

Name ______________________________ Date ______________

Lenses, Eyesight, Sources and Uses of Light Review

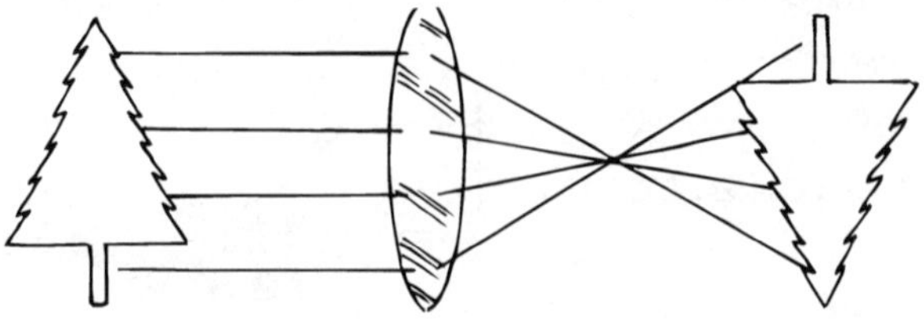

Across

1. The colored part of the eye is called the ______.
3. The ______ is an opening in the iris.
6. Nerves in the eyes take the image to your ______.
9. A ______ lens is thicker in the middle than at the edges.
10. Leaves of green plants trap energy from ______.
11. Light ______ are bent when they pass through a lens.

Down

2. Light that bounces back is called ______ light.
3. Oil was formed by rotting ______.
4. Fuel is stored ______ from the sun.
5. A ______ sends out a narrow, intense beam of light.
6. Vitamin D produces strong teeth and ______.
7. Lenses that are thicker at the edges are called ______.
8. Gasoline used in automobiles comes from ______.

Name ______________________ Date ______________

VIBRATION OF SOUND

Sound is made when something **vibrates** (moves back and forth). Sometimes the vibration is too fast or too slow for our ears to hear.

ACTIVITY

Make things vibrate to produce sounds.

1. Loop one end of a rubber band around an object such as a chair. Pull the other end a little, but not enough to break it. Pluck it. What vibrates? ______________________

   ____

   Is a sound produced? ______________________

2. Put your fingertips to your throat while you sing or speak. Move them until you feel vibration. Our windpipes have vocal cords that vibrate when air passes over them.

   Place a thin wooden or plastic ruler on a table so that about half of the ruler sticks out beyond the table's edge. Hold the ruler down firmly with one hand. With the other hand, press down on the free end of the ruler, then let it snap up.

   Can you see it vibrate? ______________________
   Does it make a sound? ______________________
   How does the sound change if you move the ruler back on the table with less of it sticking out? ______________________

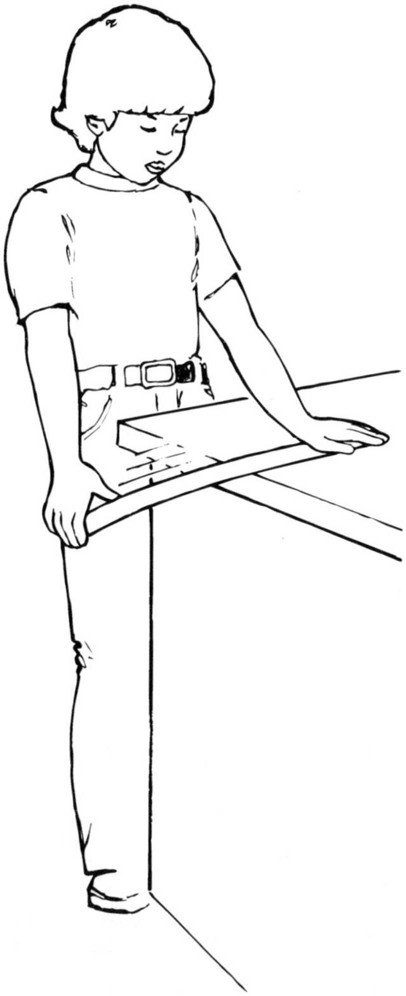

3. Strike a stainless steel fork with a spoon. You probably can't see it vibrate because it moves very little. Strike the fork again and quickly move it close to your ear. Do not touch it to your ear or hair.

   Does it sound different? ______________________
   How? ______________________

4. Strike the fork again, quickly move it close to your ear, and this time touch it to your ear or hair.

   What happens to the sound? ______________________
   What causes this? ______________________

Name ______________________________ Date ______________________

How a Sound Wave Travels

Vibration produces sound. When an object vibrates it moves back and forth. As it moves forward, it causes the air molecules near it to move away. These molecules bunch up with the next molecules to form a **compression**. When the object moves back, it leaves extra space for the molecules to spread out, forming a **rarefaction**. The distance between one compression (or rarefaction) and the next is called a **wavelength**.

Sound travels in waves. **Only the wave travels.**

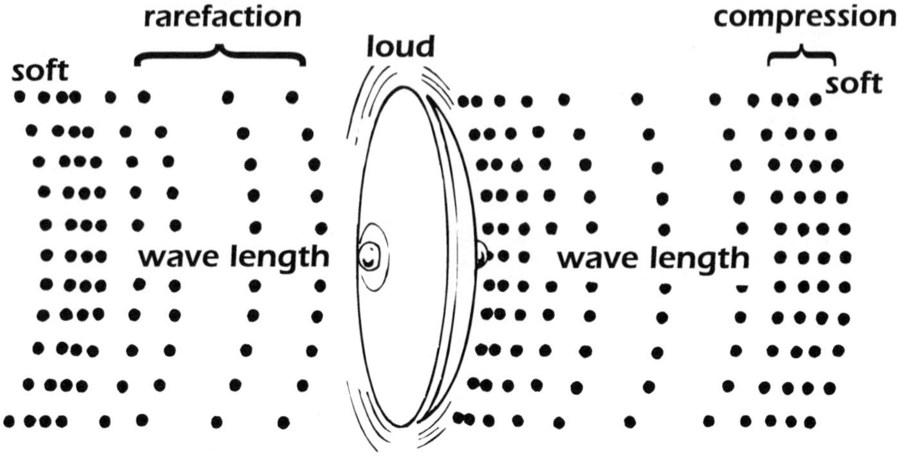

Individual molecules vibrate back and forth. They do not travel.

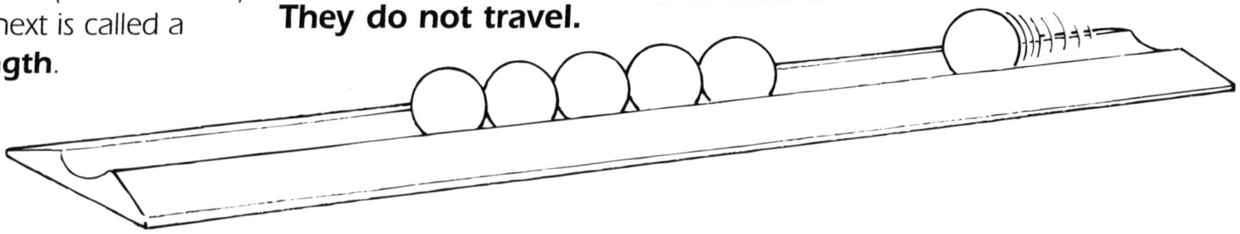

Activity

Place 5 marbles along the groove of a 12-inch ruler as shown. Now take a sixth marble and flick it against the end of the row. What happens?

____

____

The energy of the first moving marble is transferred through the other marbles to the last one. In this way, the **energy wave** travels in one direction, but the individual marbles do not.

As the wave travels, some energy is lost with each molecular compression and eventually no energy is left. That is why sounds are louder when you are closer to the source.

Molecules that are already close together are easier to compress than ones that are far apart. Therefore, since sound waves depend on compression, sound travels faster in a denser medium.

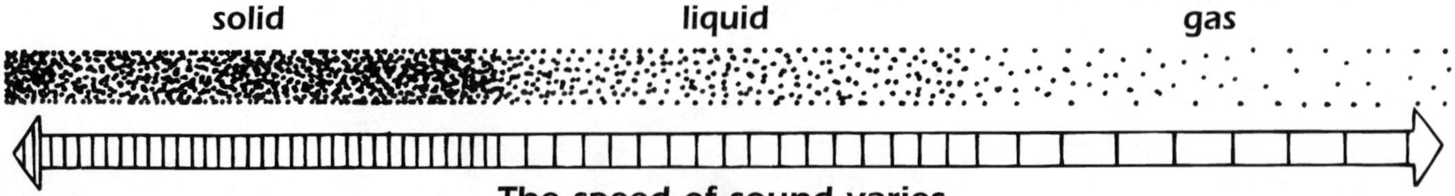

The speed of sound varies.

Name ______________________ Date ______________

A Medium for Sound Waves

Sound travels in waves. Sound waves can travel through almost anything, but they must travel through something. That something is called a **medium**. Sound waves travel faster through denser mediums. (Water is denser than air; solids are denser than liquids.)

They cannot travel, as light waves can, through completely empty space where there is no air. Light travels much faster than sound. Therefore, you can sometimes see something before the sound of it reaches you.

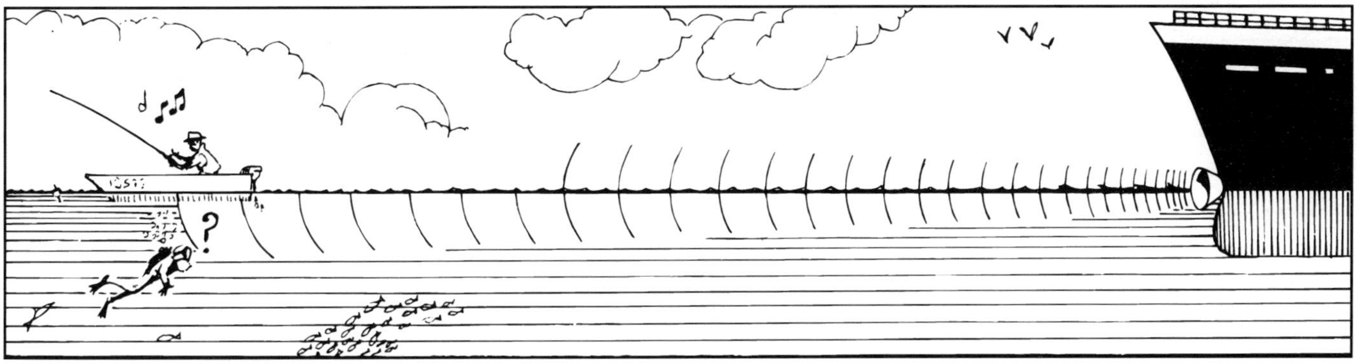

Activity

Go with a partner to a large area such as a playground when it is fairly quiet. Ask your partner to clap as soon as he or she hears your clap. Move to opposite ends of the playground. Do not look at each other. Clap your hands once loudly. Clap again when you hear your partner's clap. Keep repeating a clap as soon as each of you hears the other clap.

1. Record the time needed for 20 claps. (Use a stopwatch or a watch with a second hand.)

____

To compare the speed of sound with the speed of light, repeat 1, except this time watch your partner and make repeated movements of your hands instead of clapping. Make the movements small and simple, like the clapping motion. Time the 20 hand movements.

2. Is there any noticeable difference between the timing of the hearing distance and the seeing distance? Explain your results. ______________________________

____

Hold one end of a ruler one inch from your ear and scratch on the far end of the ruler. Notice how loud the sound is. Then move the ruler one length away and scratch on the end that is closer to you. You will be making the noise at the same distance from your ear as before.

3. Compare the sound. What do you notice? ______________________________

____

Explain this. ____

Name ______________________________ Date ______________

PRODUCING AND CONTROLLING SOUND

Producing or making simple sounds is usually easy to do. It is only through controlling sound, however, that we are able to produce complex sounds like speech or music.

ACTIVITY

Assemble several musical instruments. If possible, get people who can play them to demonstrate the sounds they make. Listen to each instrument carefully and observe how the sounds are produced. Then fill out the chart below.

Instrument	How is the Sound Produced?	What Vibrates?	How is the Pitch Changed?

CONTROLLING SOUND

Just as a mirror reflects light, some surfaces reflect sound. Sound reflection that you can hear as a separate repetition is called an **echo**. We do not hear an echo if we are close to the reflecting surface. When there are many reflecting surfaces, the echo bounces back and forth several times. This is called **reverberation**. Some substances send back few or no sound reflections. They absorb sound. Materials that absorb sound are often used in classrooms, offices, and other places to help reduce noise. Theaters and auditoriums are designed and built in ways that cause sound to travel clearly from the stage, preventing echoes. When the sound is well controlled in a room, we say it has good **acoustics**.

ACTIVITY

Place a metal pail or wastebasket close to your face and call into it. Then stuff some cloth down inside and call into it again.

1. Is the sound louder or softer with the cloth inside? ______________________________
2. Explain what makes the difference. ______________________________

Name ____________________ Date ____________

Different Ways of Producing Sound

Vibrations can be started in different ways.

vibrating strings

vibrating air column

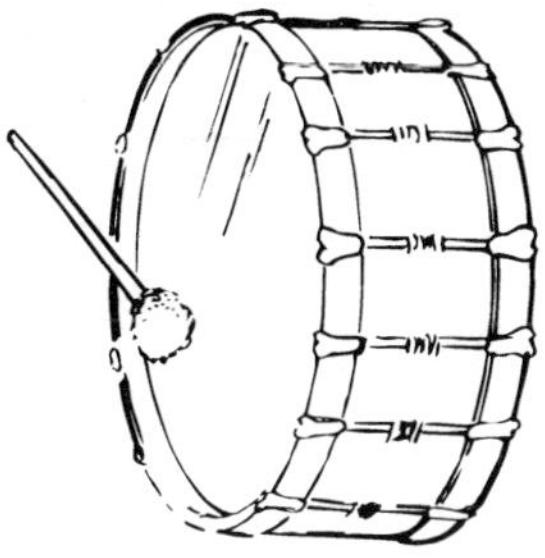

vibrating material

Activity

Draw a line from each picture to the box on the left that tells how the object produces its sound.

stroking

blowing or forcing air

plucking

hitting

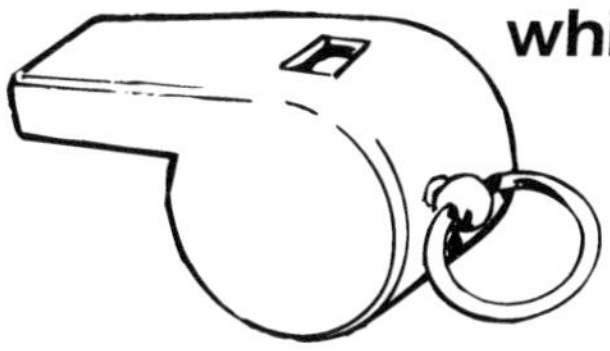

whistle

person talking

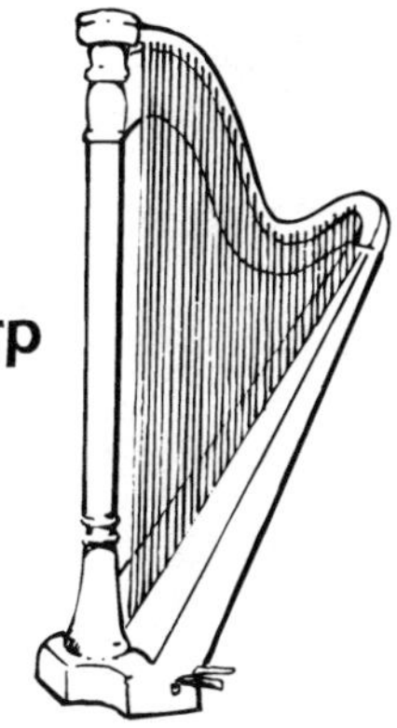

harp

piano

xylophone

cricket chirping

dog barking

Name ______________________________ Date ______________

Vibration, Sound Travel, and Producing Sound Review

Use words from the box to complete the sentences.

vibrates	reflect	waves	absorb
molecules	solids	echoes	

1. Sound is made when something ______________.
2. Sound travels in ______________.
3. Some substances send back few or no reflections. They ______________ sound.
4. Sound reflections that can be heard as separate repetitions are called ______________.
5. Sound travels faster through ______________ than through air.
6. Some materials ______________ sound waves and bounce them at an angle.

Circle the correct answer.

7. Why do you hear thunder after you see lightning?
   a) The thunder is farther away.
   b) Sound travels more slowly than light.
8. If you beat a drum on a planet where there was no air, what would happen?
   a) There would be no sound waves carried.
   b) The sound would be louder than on earth.
   c) The sound would echo.
9. Long ago, Native Americans put their ears to the ground to find out if horses were near. How did this help them?
   a) Sound cannot travel through air.
   b) The sound from the hoofbeats traveled through the ground.
   c) They could hear an echo.
10. Sounds travel faster through steel than through air. Why?
    a) Molecules are closer together in the steel.
    b) Steel is colder.
    c) Molecules in air are weaker.

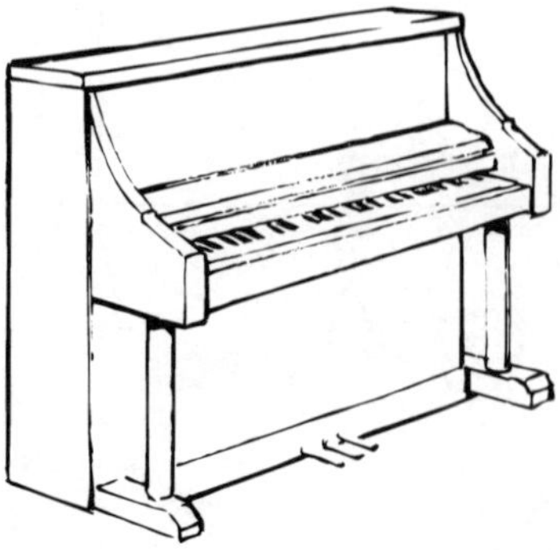

Name ______________________________ Date ______________

Characteristics of Pitch

Sometimes objects vibrate at a constant rate and produce a steady tone. The highness or lowness of a tone is called its pitch. The faster something vibrates, the higher its **pitch**. The number of times an object vibrates is called its **frequency**. The normal human ear can hear sounds with frequencies between 20 and 20,000 vibrations a second. Some animals can hear sounds under or over these frequencies.

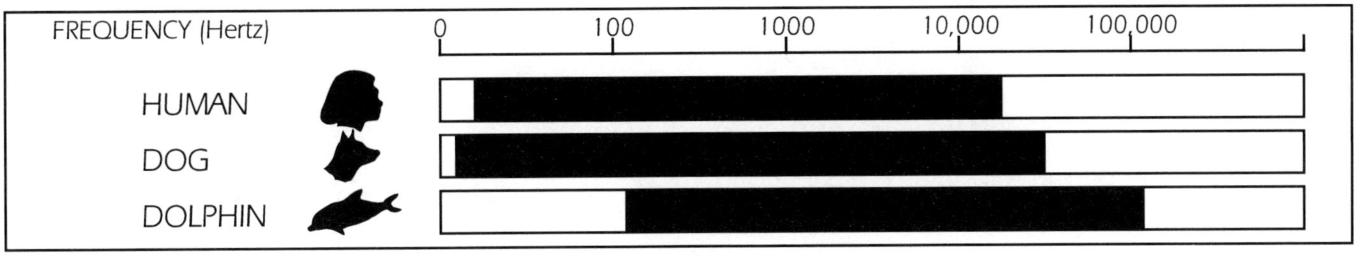

Activity

Cut a rubber band so it is no longer a loop. (a) Pull it taut (stretched tight). Have someone gently pluck the rubber band with a finger. Notice the pitch. (b) Next, shorten the length of the rubber band by holding it in the middle. Repeat the experiment.

1. Did the pitch change?______________________

   If so, in what way did it change? ______________

(a) Hold the rubber band rather loosely this time and pluck it. (b) Keep your hand at the same place on the rubber band, but stretch it tightly. (You increase the tension.)

2. What happens to the pitch?______________________

   ____

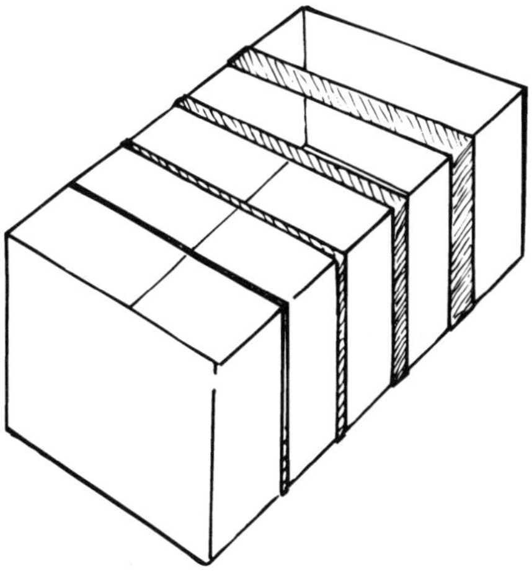

Find several rubber bands that are the same length, but different in thickness. Slide the rubber bands around an open box or pan. Pluck each one and listen to its pitch.

3. Which ones make the lowest sound? ______________

   What can you do to change the pitch of the rubber bands?______________________

   ____

Name ______________________ Date ______________

MORE CHARACTERISTICS OF PITCH

ACTIVITY

Cut a drinking straw as shown. Put the longer end in a glass of water and blow softly through the shorter end so that you produce a whistle. It may help to pinch the end of the blow-tube a little where it joins the other part that goes down in the water. Once you are able to make a sound, raise and lower the glass of water while you blow.

1. What happens to the pitch? ______________
2. When the air column is longer, is the pitch lower or higher? ______________

Get a set of bottles and a set of glasses that are the same size, shape, and material. Put increasing amounts of water in them. Strike the sides of the glasses one at a time with a spoon or other object.

3. Which ones have low pitches? ______________

   Why? ______________
   (HINT: When you add water, you add **weight** to the glass.)

Blow across the tops of the bottles to produce sounds.

4. When blowing into bottles, the bottle with the (least, most) water is lowest in pitch because the vibrating column is (longest, shortest).
5. Striking the glasses works differently than blowing into bottles. How does it work differently? ______________

   ____

6. List three conditions that change the frequency of vibration (the pitch) of a sound.

   ____

   ____

   ____

Name ______________________________ Date ______________

Loudness of Sound

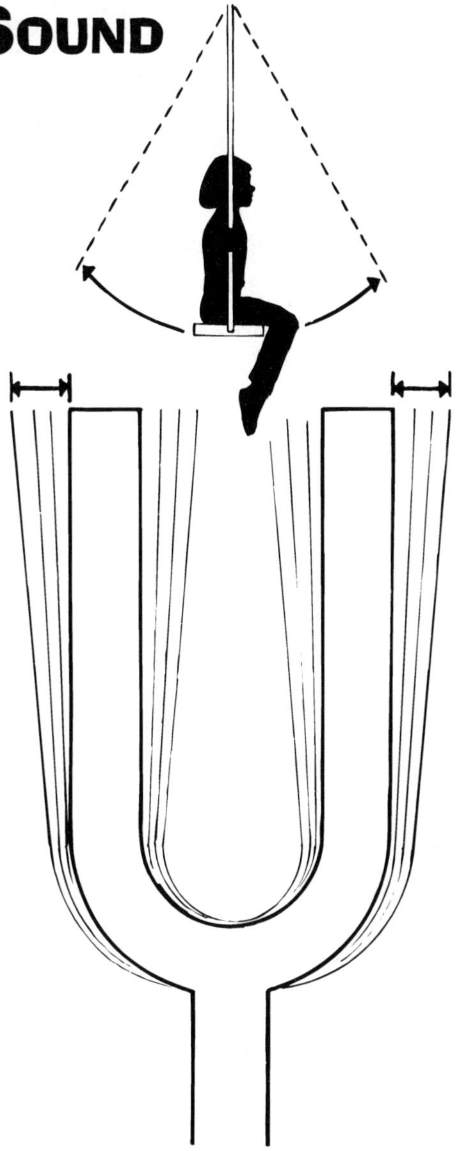

Sounds of the same pitch can be loud or soft. Several conditions affect loudness.

Distance. If you stand close to the source of a sound, the sound is louder than if you stand at a distance.

Amplitude. This is the distance that a vibrating object moves as it vibrates. Think of yourself on a swing. The distance you move forward and backward would be like amplitude. It takes more energy to produce a large amplitude.

Density. Dense materials have molecules that are close together. Water is denser than air. Strike two stones together. Listen to the sound. Then submerge the stones in a pan of water, strike them, and listen to the sound they make.

1. Why is the sound louder in the water?

2. Usually sound travels (faster, slower) in denser materials. The sound is louder because it reaches your ears sooner.

Activity

Change the amplitude of vibrating objects and notice the change in loudness. Pluck a rubber band or string gently, and then with more energy. Tap a table with a ruler lightly, then hard. Write SOFTER or LOUDER on the lines to show results.

3. ______________________ 4. ______________________

(rubber band)

gentle plucking (small amplitude) | forceful plucking (large amplitude)

5. ______________________ 6. ______________________

(tapping ruler)

light tapping (small amplitude) | hard tapping (large amplitude)

Name ______________________________ Date ______________

MORE WAYS TO AFFECT LOUDNESS

The size of the vibrating surface affects the loudness of a sound.

ACTIVITY

Strike the prongs of a metal fork in mid-air with a spoon. (Make sure the fork is not too thin or light-weight.) Listen to the sound. Repeat, but this time quickly press the handle of the fork to a table. Notice the difference in loudness.

1. Why is the sound louder when the fork is on the table? ______________________________

____

____

Run your finger along the teeth of a comb held in your hand. Then try it again while you hold the end of the comb tightly against a window, a door, a wooden table, or other large object. If you can make a large surface vibrate, it will send out more energy to the air that touches it.

2. Will the sound be louder or softer? Try a variety of surfaces. Do hard or soft surfaces make the sound louder?

____

You can increase your ability to hear sounds by catching more of the sound waves that approach you. Cup your hands around your ears and listen to a sound. Take your hands away and listen. Then put your hands in front of your ears, palms facing backwards, and listen.

3. Which way made the sounds loudest? ______________________________

   Which way made the sounds hardest to hear? ______________________________

   Explain the difference. ______________________________

Have you ever noticed how some animals rotate their ears to catch sounds?

Tie a string to a metal spoon and hold the string to your ear. Hit the spoon with another spoon and hear it ring. Hold the string away from your ear and strike it.

4. Which way sounds loudest? ______________________________

____

Some people who have a hearing loss can still detect vibration through the bones of their skulls. Hearing aids amplify vibrations and send them through these bones.

Name ______________________ Date ______________

LIGHT & SOUND

Characteristics of Sound

Frequency and **amplitude** are the scientific words used to describe pitch or tone and loudness. Remember: the greater the frequency, the higher the pitch of the sound; and the greater the amplitude, the louder the sound.

1. Arrange the sound waves from loudest to softest: ___ ___ ___ ___
2. Arrange the sound waves from highest to lowest: ___ ___ ___ ___

A
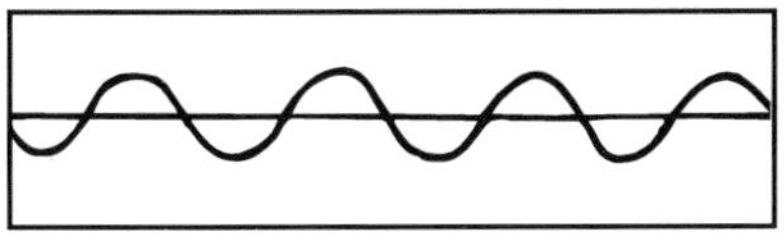
low amplitude

B
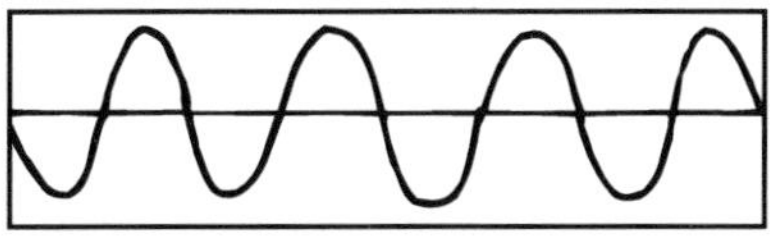
high amplitude (sound more intense)

C
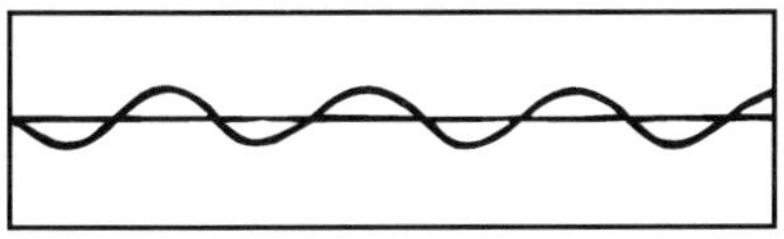
low pitch

D
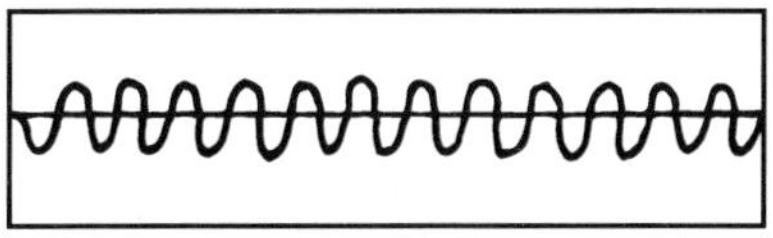
high pitch (rapid frequency)

Decibels—Measure of Sound Intensity

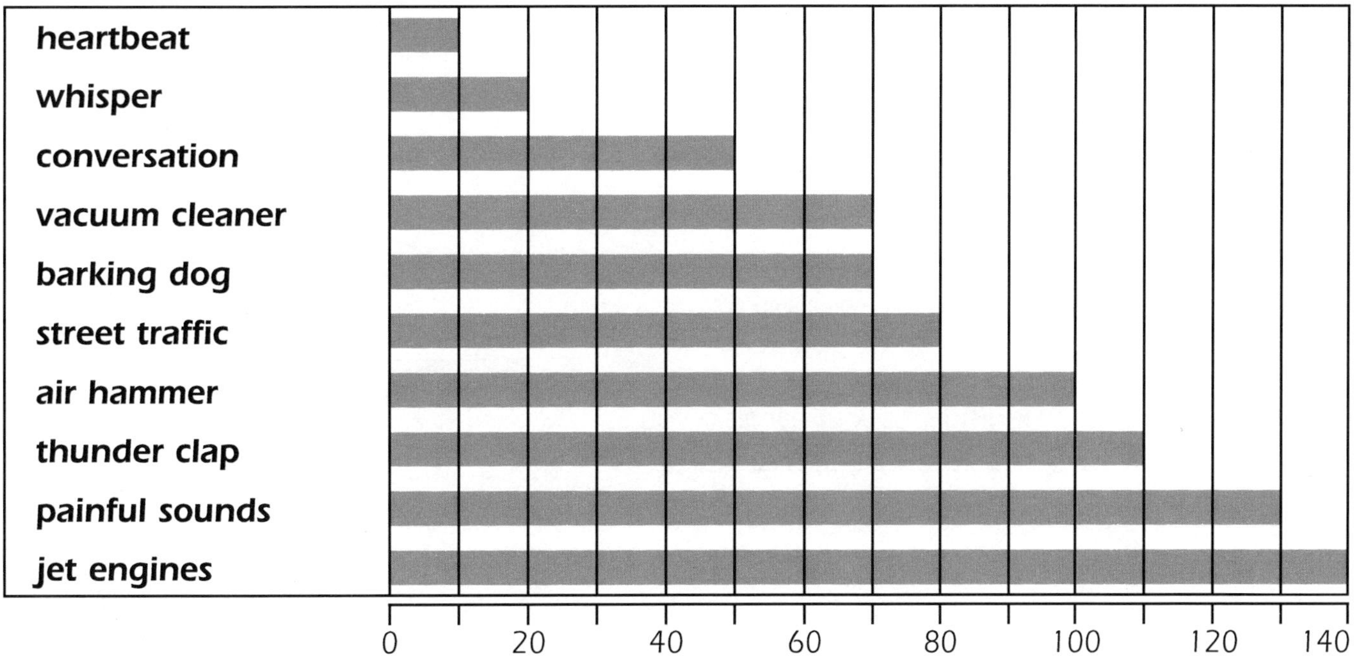

Compare the sounds below with those on the chart. Estimate what the sound intensity might be for each.

a cat meowing __________ Niagara Falls __________ a computer keyboard __________

a video game __________ singing __________ a pin dropping __________

Name ______________________________ Date ____________________

Pitch and Loudness Review

Write the word that correctly completes each sentence.

1. More rapid vibrations give ____________________ pitch.
   higher lower louder

2. Pitch is changed by the ____________________ of vibration.
   length size frequency

3. ____________________ does not affect pitch.
   weight length amplitude

4. Pitch becomes ____________________ when tension increases.
   higher lower louder

5. Loudness is increased with greater ____________________.
   amplitude molecules tension

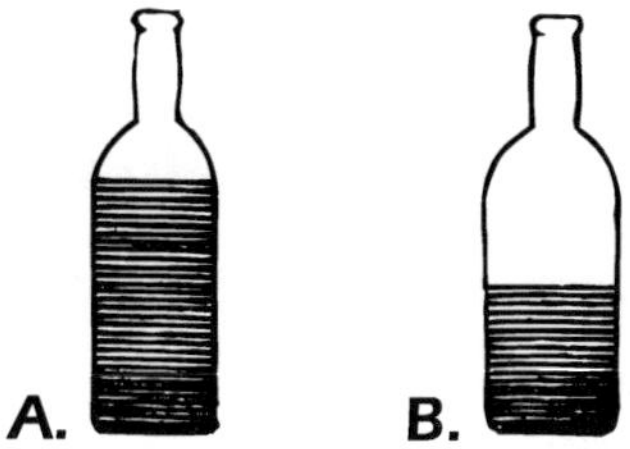

Answer the following questions.

6. Which bottle will have a lower pitch? __________
   Why? ______________________________

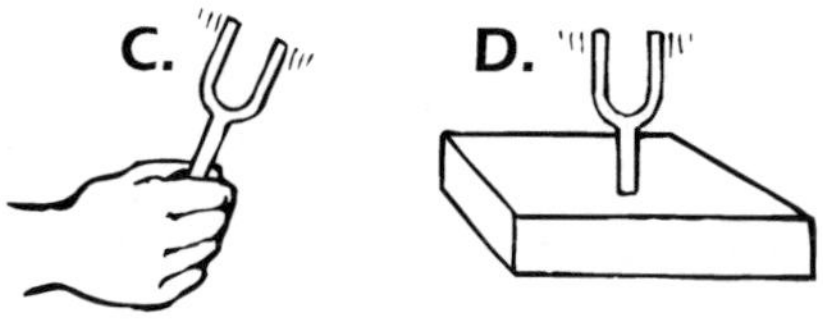

7. Which vibrating fork will sound louder? __________
   Why? ______________________________

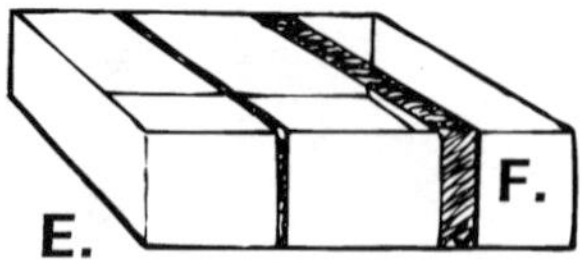

8. Which rubber band will have a lower pitch? ____
   Why? ______________________________

9. Why will this cone held against your ear help you hear better? ______________________________

10. Which animal sound will travel faster? __________
    Why? ______________________________

Name ______________________ Date ______________

LIGHT & SOUND

Sound and Hearing

Our ears are sensitive to the vibrations of sound. The **pinnacle**, the outer part of the ear we can see, catches the sound and funnels it into the **ear canal**. Sound waves striking the eardrum cause it to vibrate, which then causes the tiny middle ear bones to vibrate, too.

The **hammer**, the **anvil**, and the **stirrup** are the smallest bones in the body. Their very small size (compared to the eardrum) amplifies the energy of the sound waves before they reach the **cochlea**. The snail-shaped cochlea converts the mechanical energy of sound vibration into an electric nerve impulse which our brain understands. The nerve carrying these messages is called the **auditory nerve**.

Our ears also help us keep our balance. The **semicircular canals** contain fluid. When we move, the fluid moves too, and this lets us know our position.

The **Eustachian** (you-**sta**-tion) **tube** connects the middle ear with the back of the mouth. When your ear pops, that's the Eustachian tube working to keep the air pressure on both sides of the eardrum equal.

Use reference sources to label the parts of the ear.

The Human Ear

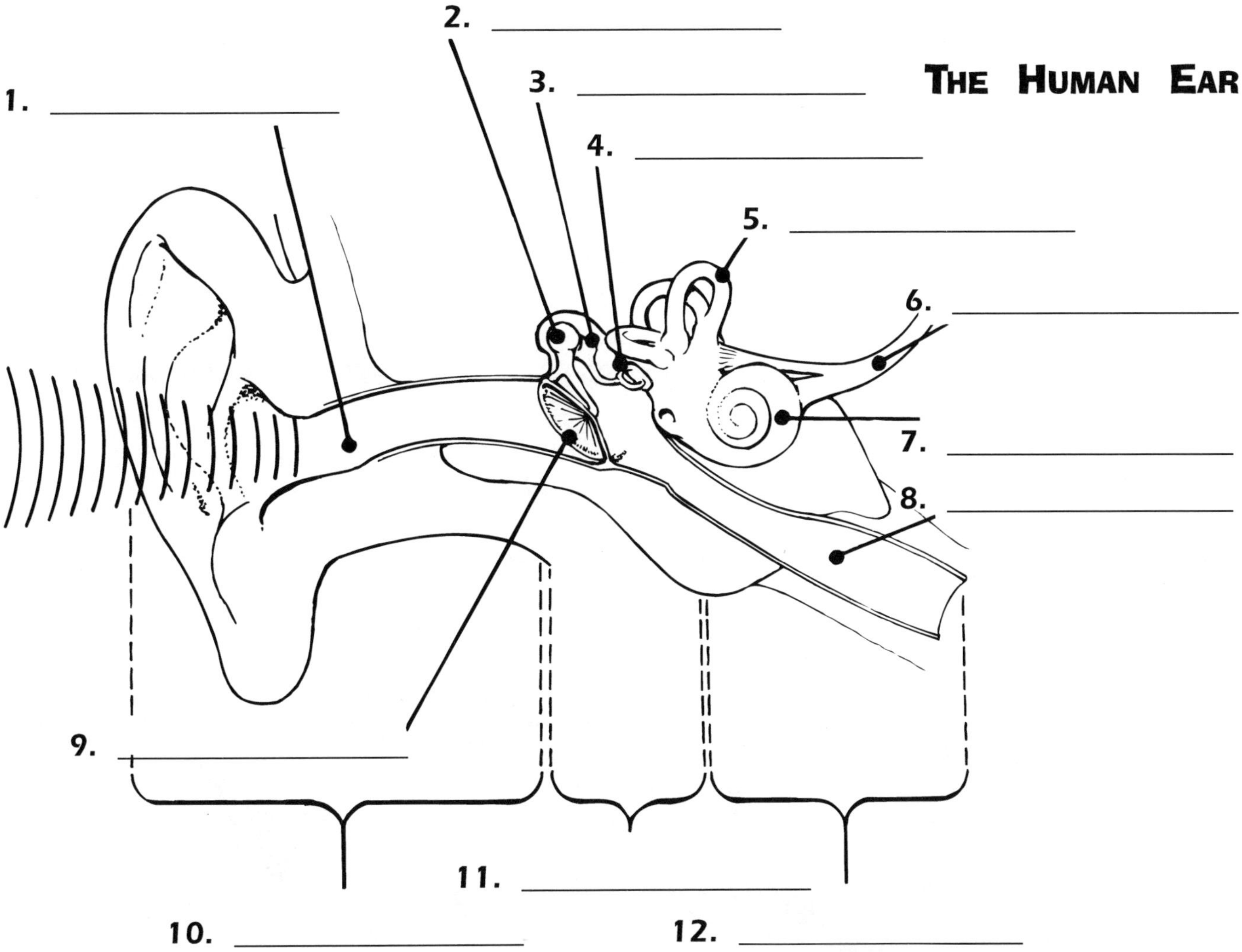

Name ______________________________ Date ______________

Using and Reproducing Sound

We use sound for many purposes, such as communications, entertainment, and warning devices.

1. List several different methods and devices that use sound to send and receive messages:

____

____

____

2. List several different methods and devices that use sound to provide entertainment:

____

____

____

3. List several different methods and devices that use sound to provide warnings:

____

____

____

Phonograph records were formerly used to save and reproduce music and other sounds. Most phonograph records have been replaced by tapes and compact discs.

Tapes and compact discs record sound using magnetism. Each sound is broken down in thousands of components (such as frequency and amplitude) that can each be defined with a magnetic code. This is called **digital recording**. On a compact disc, this code appears as tiny uniform peaks and valleys. A laser beam acts like the needle of a phonograph; it moves over the disc and reads the peaks and valleys.

Although digital recording has revolutionized the recording industry, three instruments are still important in sound recording: the **microphone**, the **amplifier**, and the **loudspeaker**. A microphone changes sound waves into an electric current. A loudspeaker changes the electric current back into sound. Most sound systems have an amplifier that changes weak currents into strong ones.

Light & Sound Background Material

Pages 1–6: Scientists think light travels through space in waves. Though such waves have not been seen, they are often depicted in diagrams. In such diagrams, the y-axis is a measure of amplitude which, in the case of light, may be likened to "brightness" or strength. The x-axis measures time. Thus, higher frequencies are shown with the waves closer together on the x-axis. Amplitude changes do not indicate that the waves or photons are moving up and down, but rather indicate that the wave is stronger in some places than others.

In some ways, light also behaves like streams of tiny particles. These particles are bundles of energy called **photons**, which are given off from the light source.

Light travels through space at 300,000 kilometers (1186,000 mi.) per second; through water at 224,000 kilometers (139,000 mi.) per second; and through glass at 200,000 kilometers (124,000 mi.) per second.

In 1860, James Clerk Maxwell developed the **electromagnetic theory of light**. This theory stated that light waves were electrical in nature, instead of mechanical, as sound waves are. Electromagnetic waves can be arranged on a graph according to their different lengths (see page 2).

Our eyes are sensitive to a certain range of electromagnetic energy. They are sensitive to the wavelengths that are most abundant in sunlight. The part of sunlight which we can see is called **visible light**. Our eyes are sensitive to only about three percent of all electromagnetic waves.

Use the illustrations on pp. 2, 3, and 5 to stimulate discussions and questions with the students. The following concepts may be of interest to students:

1) **Cosmic rays** are particles of high energy that originate in outer space. They crash into atoms in the air and create secondary particles which reach Earth. A human being can tolerate a minute dose of cosmic radiation each day without damage. However, at extremely high altitudes or beyond the atmosphere, the dosage is probably hazardous. This presents a problem for space travelers. Cosmic rays were discovered in the early 1900s.

2) The **gamma-ray** is similar to the x-ray but has a shorter wavelength. Gamma rays are given off by radioactive elements such as radium. Large amounts are harmful to the body, but small amounts can be beneficial. They are used in x-ray machines to examine the body and to treat some diseases.

3) There are many medical uses of **x-rays**, for examination and treatment. Science also has a variety of uses for x-rays. In industry, x-rays can be used to examine objects for flaws.

4) **Ultraviolet rays** are sometimes known as **black light**. They cannot be seen by the human eye. The sun is a source of ultraviolet rays, but the air absorbs much of this radiation. Some chemicals become **fluorescent** (give off visible light) when placed under ultraviolet rays. This fluorescence has industrial uses (identifying ores, lighting, etc.). Certain ultraviolet rays can kill cells, help sterilize water or air, and treat diseases. Milk can be treated with ultraviolet rays which cause the formation of vitamin D within the body.

5) **Infrared rays** have wave-lengths too long to be seen by the human eye. These rays are given off by hot objects. They are able to pass through materials that block ordinary light. Infrared rays may be used to treat skin diseases and to dry paint.

Pages 7–10: Reflection is the bouncing back of a ray of light. Some surfaces reflect better than others. Rough surfaces scatter light rays at mixed angles so reflection is poor. This is termed **diffuse reflection**.

Smooth, shiny surfaces, however, reflect light well. Surfaces like mirrors reflect light without scattering it, and produce **regular reflection**.

The angle at which a light ray strikes a surface is equal to the angle at which it leaves when reflected back. The **angle of incidence** is formed by the line of the light ray as it hits the surface and a line perpendicular to the surface at that point. The **angle of reflection** is formed by the same perpendicular line and the ray of reflected light. These two angles are equal.

Flat mirrors reflect light in the same pattern as the

rays strike the mirror. There are curved mirrors which have special uses. A mirror that curves to the inside reflects light rays so that they come together in one spot. A reflector such as this is used in automobile headlights to focus light on the road. Some mirrors curve outside to spread the light rays. These mirrors make it possible to see a wider area and are used as rear-view mirrors.

Pages 11–14: Objects that make their own light are described as being **luminous**. We see nonluminous things by **reflected** light.

The sun is our best source of light. It gives heat as well as light, and it is the original source of energy on our planet. Some light is also produced from burning fuel.

Electric light bulbs use a coiled metal wire called a **filament**. Air inside the bulb is replaced by nitrogen gas, which causes the filament to glow. Light bulbs give off 25% light and 75% heat.

Neon lights are tubes filled with neon gas. An electric current is passed through the tube and makes the gas glow.

Fluorescent light uses tubes coated on the inside with a chemical called **phosphor**. It glows when struck by ultraviolet rays. These rays are formed by an electric current flowing through mercury gas in the tube.

Fireflies and certain fish and bacteria produce a cold light. They give off light without producing any heat.

Light has many technological applications. Lasers, (**L**ight **A**mplification by **S**timulated **E**mission of **R**adiation) alone have dozens of uses in surgery, communications, and space technology. It is good to emphasize to students that other light waves—radio and microwaves, for example—are the same energy form as the visible light waves even though we cannot see them.

Pages 15–17: Sound is a form of energy that is produced by a vibrating object. When an object vibrates, it makes the surrounding air vibrate. The object's outward vibration compresses the air. The object pulls away from the air on the other side and leaves a space of thinner air. More air rushes into this space, which leaves thinner air a little farther away. Each place where the air molecules are pressed together is called a **compression**. Between the compressions, the molecules spread apart and are called **rarefactions**. The distance from the middle of one compression to the middle of the next is called a **wavelength**.

Each molecule moves only a little. It is pushed out of place by another molecule and then springs back to its original position. This motion goes on in waves of many molecules. As the waves move farther away from the vibrating object, the sound energy is used up, and the sound ends. Thus, it is the energy of sound that travels, not the individual molecules. This is what is meant by "mechanical" wave.

Compare sound waves with light waves. A similar diagram is used to represent sound waves, but here the amplitude does show actual physical displacement of the particles.

Sound cannot travel in a vacuum. It must travel through some substance. Sound travels through some materials faster than through others. Molecules are closer together in solids than in liquids; they are closer in liquids than in gases. The closer together the molecules are, the faster the sound travels through them. In general, sound travels faster in denser materials.

Temperature affects the speed of sound. Molecules move faster in warm air than in cold. Sound is carried faster, then, in warm air.

Airplanes can now fly faster than the speed of sound. The airplane sends out sound waves in front of it, then catches up with these waves and finally pushes them together. It must break the barrier of waves it has squeezed together. When it does this (at around 760 mph), it creates a shock wave heard as a loud sonic boom.

Pages 18–20: Discuss the various methods of producing sound that are illustrated on page 19. Discuss what vibrates in each one.

Demonstrate these methods by using simple everyday objects such as plucking a rubber band, hitting a glass with a pencil, stroking sandpaper across materials, or blowing across a soda bottle. Discuss what vibrates to produce each sound.

Sound that bounces back is called an **echo**. Echoes may be heard from reflecting surfaces if one is standing a certain distance away. Echoes are fainter than the original sound because energy is lost during its return. The reflection of sound is like the reflection of light: when a sound wave strikes a reflecting surface at an angle, it is reflected at exactly the same angle.

Pages 21–26: Pitch is controlled by the number of vibrations per second made by a sound-producing object. The number of vibrations per second is the frequency. The higher the frequency, the higher the pitch.

Several factors can affect vibrating frequency and pitch: 1) Length: short lengths vibrate faster than long lengths; 2) Tension: tightly stretched strings vibrate faster than loosely stretched strings; 3) Width/Weight: narrow strings vibrate faster than thick strings of the same material.

Ultrasound refers to frequencies too high to be heard by humans—more than 20,000 vibrations per second. Ultrasonic waves have many uses: to detect flaws in metal, to kill insects, to pasteurize milk, to clean surgical instruments, and to treat diseases.

Bats, and some sea animals such as the porpoise, use **echo-location** to help them locate prey. Bats produce high-pitched sounds that bounce back when they strike an insect or obstacle. The bat then follows the path of the echoes.

Radar (**RA**dio **D**etecting **A**nd **R**anging) is a device that sends out short, intense pulses of radio waves and listens between the pulses for echoes (reflected radio waves bounced back from objects).

Sonar (**SO**und **N**avigation **A**nd **R**anging) is a detecting device that uses sound. It sends out a sound wave into the water. The wave is reflected when it meets an object. The distance to the object is measured by the time taken for the wave to return. Sonar is used to determine the depth of water. Fishermen use sonar to locate fish.

An instrument called a **seismograph** records vibrations in the earth. Geologists often use a seismograph to locate minerals. Small explosions are detected and the vibrations recorded. Geologists can tell through what kinds of rock the vibrations passed, and from what kinds they were reflected. They also know echo patterns that indicate the presence of oil or other minerals.

Substances that send back few or no sound reflections **absorb** sound. Such materials are used in construction to reduce building noise levels.

Loudness of sound depends upon several variables: 1) Loudness decreases as the distance increases between the receiver and the source of the sound; 2) **Amplitude** is the distance that a vibrating object moves as it vibrates. The greater the amplitude, the louder the sound; 3) Loudness decreases as density of the transmitting material increases; and 4) Loudness increases as the area of vibration increases.

The **intensity** of a sound (the amount of energy carried by the sound waves) is measured by a unit called the **decibel**. Decibels are measured on an instrument called a sound-level meter, which changes sound energy to electrical energy. The sound energy can then be read on the meter.

Pages 22–28: The human ear is divided into three main parts: outer ear, middle ear, and inner ear. The structure of the ear allows sound vibrations to pass from outside the head to the brain, where the sounds are interpreted. The outer ear catches the sound waves. The sound waves make the eardrum vibrate. The vibrating eardrum then makes three tiny bones (ossicles) vibrate. (The ossicles are the three smallest bones in the body.) Vibrations pass on to the cochlea. Inside the cochlea, sound waves are turned into nerve impulses. The nerves send messages through the auditory nerve to the brain.

Some noises can be harmful to the ear. It has been shown that people who live and work in very noisy places experience gradual hearing loss. Safety regulations regarding sound levels are designed to protect people. Eighty-five decibels is recommended as the loudest permissible level.

Loud music can cause hearing loss because high-powered amplifiers can make the sound near the loudspeakers so intense that, with repeated exposure, damage to ears may occur.

The basic principle of hearing is applied when recording sound: the mechanical energy of the sound wave is changed into electrical energy, which is either interpreted or stored. As with the camera/eye analogy, the ear can be likened to a recording system: the microphone, like the cochlea, converts the mechanical energy to electrical; the amplifier, like the ossicles, makes a weak signal stronger; and the auditory nerve is the "wire" that carries the signal where it can be stored.

ANSWERS

Page 1

1. waves
2. light bulb
3. second
4. straight

Page 2

1. no; Light travels in straight lines.
3. no
4. Clouds reflect the light.

Page 3

The pencil looks broken.

Page 4

Transparent: water, air; Translucent: waxed paper, paper, cotton cloth, person's hand; Opaque: aluminum foil, wool cloth, cardboard, wood.

1. Shadows get larger when moving back.
2. getting closer makes the shadows smaller.

Page 5

1. Spectrum colors: red, orange, yellow, green, blue, violet.
2. Objects that reflect no light are black.
3. White paper reflects all colors.
4. Dark objects absorb more than lighter objects.

Page 6

1. radiant
2. straight
3. transparent
4. refracted
5. opaque
6. spectrum
5. opaque
6. spectrum
7. reflected
8. violet
9. visible
10. absorbs

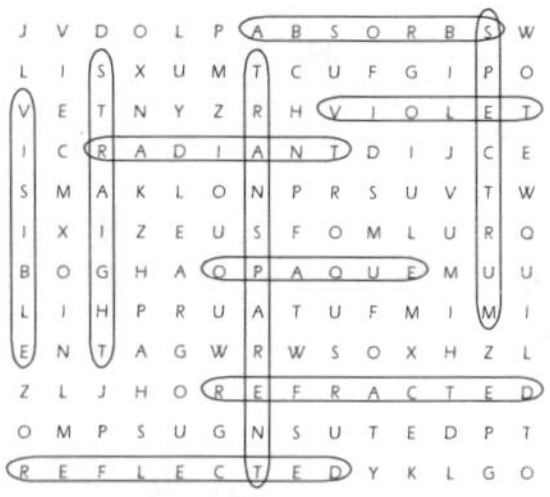

Page 7

1. The arrow points the opposite way.
2. It changes direction when the glass is in front.
3. The glass of water acts like a lens.
4. convex
5. concave

Page 8

The pupils become very large in a darkened room to let in more light.

Page 9

1. lens
2. film
3. brain
4. no
5. image

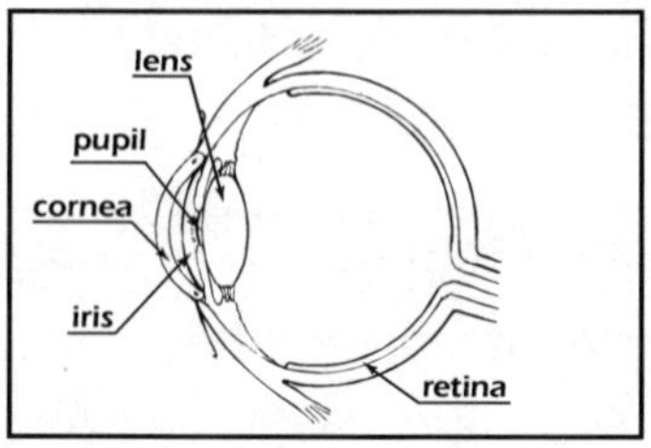

Page 10

1. Light traveled in a straight line because you can see your image.
2. You can see the pencil because it is reflected at the same angle.
3. You cannot see the pencil if you put it on C.

Page 11

1. Natural light: moon, firefly, deep-sea fish
   Artificial light: incandescent light bulb, fluorescent light bulb, neon sign, candle
2. firefly
3. naturally (as with the sun and firefly) or artificially (as in electric light bulbs or candles)
4. light bulb, lamp, sign, candle
5. incandescent light bulb, fluorescent light bulb, neon sign, candle

Page 12

1. Stored energy becomes fuel and keeps the earth's water cycle working. Ultraviolet light helps the body make vitamin D.
2. Coal and oil come from ancient rotting plants.
3. Leaves of green plants trap energy.
4. Energy in meat comes from the plants that those animals ate.

Page 13

1. Solar cells, photography, treating eye problems, transmitting phone, radio, and television signals, microwave cooking, storing information (compact disc).
2. A solar cell uses sunlight to generate a small electric current.
3. **L**ight **A**mplification by **S**timulated **E**mission of **R**adiation.
4. **RA**dio **D**etecting **A**nd **R**anging.

Page 14

Across		Down	
1.	iris	2.	reflected
3.	pupil	3.	plants
6.	across	4.	energy
9.	convex	5.	laser
10.	sunlight	6.	bones
11.	rays	7.	concave
		8.	oil

Page 15

1. The rubber band, yes
2. yes, yes, with less sticking out, the sound is higher.
3. yes, the sound is louder
4. The sound stops because the vibration stops.

Page 16

The energy of the first marble is transferred through the other marbles to the last one.

Page 17

1. Answers will vary.
2. Time is shorter the second time because light travels faster than sound.
3. Sound is louder the first way; the ruler carries the sound better than air.

Page 18

1. Sound is softer with cloth inside.
2. Cloth absorbs the sound.

Page 19

whistle: blowing; harp: plucking; piano: combination of hitting/plucking; xylophone: hitting; person talking: forcing air; cricket chirping: stroking; dog barking: forcing air.

Page 20

1. vibrates	6. reflect
2. waves	7. b
3. absorb	8. a
4. echoes	9. b
5. solids	10. a

Page 21

1. yes, it became higher
2. The pitch gets higher.
3. The widest rubber band makes the lowest sound. Pulling it tighter (increasing tension) changes pitch.

Page 22

1. The pitch changes.
2. It is lower when the air column is longer.
3. Those that contain the most water have the lowest pitch because they are the heaviest.
4. Least, longest.
5. The bottles produce sound in the air column, the glasses produce sound on the side of the glass.
6. weight, length, tension.

Page 23

1. Water is denser than air.
2. faster
3. softer
4. louder
5. softer
6. louder

Page 24

1. The table increases the size of the vibrating object.
2. Louder; the sound is louder on a hard surface.
3. Palms cupped forward, palms backwards; palms cupped forward.
4. Louder when touching the ear.

Page 25

1. B, D, A, C
2. D, B, A, C

Page 26

1. higher
2. frequency
3. amplitude
4. higher
5. amplitude
6. B–Longer vibrating column
7. D–the box increases the size of vibrating object
8. F–it is heavier
9. It catches more sound waves.
10. H–sounds travel faster in water than in air.

Page 27

1. ear canal	7. cochlea
2. hammer	8. Eustachian tube
3. anvil	9. eardrum
4. stirrup	10. outer ear
5. semi-circular canal	11. middle ear
6. auditory nerve	12. inner ear

Page 28

Answers will vary.